The Mean

John Fiske

Alpha Editions

This edition published in 2022

ISBN : 9789356895850

Design and Setting By
Alpha Editions
www.alphaedis.com
Email - info@alphaedis.com

Contents

INTRODUCTION

The new significance of education

The last century has witnessed an unprecedented development in the significance of education. One direct consequence has been an increased reverence for childhood. In this movement which has increased the dignity of children and schools, two large forces have been at work,—one social and the other scientific. The growth of the democratic spirit among men and institutions has made the education of children a public necessity, and lifted the school to a position of high social importance. The application of the theory of evolution to man and his life has revealed human infancy as one of the largest factors making for the superiority of man in the struggle for existence, and given to childhood a vast biological importance. The necessities of democracy and the truths of science, acting more or less independently of each other, have given to education a breadth of meaning which it did not possess before. They have shown that infancy is the largest opportunity and education the most powerful instrument for the conscious adjustment of man to the physical and social world in which he lives.

Democracy changes the function of schools

It was the attempt of democracy to educate all of its children which was the initial and important event that provoked large changes in our notions of the social function of education. As long as the school was for the few, and such it was in the less liberal periods of history, the school tended to be an authoritative institution with more or less rigid methods of procedure. With fixed ideas of truth and the means of acquiring truth, it was to a considerable degree unbending in its attitude toward youth. Even if freedom from economic toil and social regulation permitted, only the type of mind that could fit the school's established institutional ways could endure its discipline and achieve its rewards. Other types of mentality it would not receive or retain as students. Under such an organization the school was selective of a special kind of talent. It was not an instrument, so adjustable in its methods of appeal and instruction, that every manner of child could gain considerable of the wisdom of the world. But when a more democratic order was established, the function of the school underwent a considerable change. Democracy granted to all men freedom in manhood; to safeguard its privileges, it had to educate all men in childhood. The school for selected scholars had to be transformed into a school for every variety of citizen. With every child sent to school by order of the state, the teacher had to forego his traditional aloofness, and to adjust his methods of teaching so that every member of the enlarged school community could come into a knowledge of the civilization in which he lived. With the inclusion of the blind, the deaf, the slow of mind, and the restless of spirit,—individuals left out of the old scheme of education and now reverently educated by the new democratic order in spite of all their defects,—the school becomes more flexible and variable in its methods of transmitting truth. More of the knowledge of human life is brought within the comprehension of children; more men are brought into a large and sympathetic participation in the activities of our civilization. In the truest sense the school becomes an instrument of adjustment between childhood and society.

Evolutionary thought interprets childhood

If the democratic movement emphasized the factor of social adjustment in the school's function, it was the scientific movement of the last half-century which drew attention to infancy as a superior opportunity for biological adjustment Among all the contributions of modern evolutionary science to educational thought, none is, more striking or more far-reaching in its implications than that special group of generalizations which states the biological function of a prolonged infancy in man. Interpreting this period, of helplessness and dependence as one of plasticity and opportunity, it has shown that the greater power of man in adjusting himself to the complex conditions of life is due to his educability, which in turn is the outcome of his lengthened childhood. This "doctrine of the meaning of infancy," for such it has been called, is perhaps best known to the teaching profession through those enlargements and applications of the doctrine which have been made by Mr. Nicholas Murray Butler in his exposition of "the meaning of education." As a belief, it is at least as old as the period of the ancient Greek philosopher, Anaximander. As a doctrine in our modern thought, it owes its influential reappearance to certain evolutionary hypotheses of Mr. Alfred Russel Wallace, which in turn stimulated Mr. John Fiske to that further inquiry which resulted in those first cogent and extended statements of the doctrine which have been the basis of so many subsequent educational applications.

Mr. Fiske's presentation of the meaning of infancy

Because of the fundamental importance of Mr. Fiske's presentation of "the doctrine of the meaning of infancy," his views are here reprinted in detail. The material consists of an essay and an address. The first of these, "The Meaning of Infancy," is a brief and simplified restatement of those theories of man's origin and destiny as first suggested in his lectures at Harvard University in 1871, and later developed more fully in the "Outlines of Cosmic Philosophy," part II, chapters xvi, xxi, and xxii. The second of these, "The Part played by Infancy in the Evolution of Man," is an address delivered by Mr. Fiske as the guest of honor at a dinner at the Aldine Club, New York,

May 13, 1895. Together these two papers constitute the most detailed and valuable elucidation of the doctrine that we possess. In offering them to the teaching profession and the reading public in this form, it is with the sincere hope that this biological interpretation of childhood and education will lend a new spiritual dignity to the whole institution of education. It must certainly be gratifying to those who are profound believers in the efficacy of education, to note that its significance is wider than its service to particular persons and states; that education is, in truth, the conscious and latest mode of that wider world-evolution which has been in progress since the beginning of time.

I

THE MEANING OF INFANCY

What is the Meaning of Infancy? What is the meaning of the fact that man is born into the world more helpless than any other creature, and needs for a much longer season than any other living thing the tender care and wise counsel of his elders? It is one of the most familiar of facts that man alone among animals, exhibits a capacity for progress. That man is widely different from other animals in the length of his adolescence and the utter helplessness of his babyhood, is an equally familiar fact. Now between these two commonplace facts is there any connection? Is it a mere accident that the creature which is distinguished as progressive should also be distinguished as coming slowly to maturity, or is there a reason lying deep down in the nature of things why this should be so? I think it can be shown, with very few words, that between these two facts there is a connection that is deeply in-wrought with the processes by which life has been evolved upon the earth. It can be shown that man's progressiveness and the length of his infancy are but two sides of one and the same fact; and in showing this, still more will appear. It will appear that it was the lengthening of infancy which ages ago gradually converted our forefathers from brute creatures into human creatures. It is babyhood that has made man what he is. The simple unaided operation of natural selection could never have resulted in the origination of the human race. Natural selection might have gone on forever improving the breed of the highest animal in many ways, but it could never *unaided* have started the process of civilization or have given to man those peculiar attributes in virtue of which it has been well said that the difference between him and the highest of apes immeasurably transcends in value the difference between an ape and a blade of grass. In order to bring about that wonderful event, the Creation of Man, natural selection had to call in the aid of other agencies, and the chief of these agencies was the gradual lengthening of babyhood.

Such is the point which I wish to illustrate in few words, and to indicate some of its bearings on the history of human progress. Let us first observe what it was then lengthened the infancy of the highest animal, for then we shall be the better able to understand the character of the prodigious effects which this infancy has wrought. A few familiar facts concerning the method in which men learn how to do things will help us here.

When we begin to learn to play the piano, we have to devote much time and thought to the adjustment and movement of our fingers and to the interpretation of the vast and complicated multitude of symbols which make up the printed page of music that stands before us. For a long time, therefore, our attempts are feeble and stammering and they require the full concentrated power of the mind. Yet a trained pianist will play a new piece of music at sight, and perhaps have so much attention to spare that he can talk with you at the same time. What an enormous number of mental acquisitions have in this case become almost instinctive or automatic! It is just so in learning a foreign language, and it was just the same when in childhood we learned to walk, to talk, and to write. It is just the same, too, in learning to think about abstruse subjects. What at first strains the attention to the utmost, and often wearies us, comes at last to be done without effort and almost unconsciously. Great minds thus travel over vast fields of thought with an ease of which they are themselves unaware. Dr. Nathaniel Bowditch once said that in translating the "Mecanique Celeste," he had come upon formulas which Laplace introduced with the word "obviously," where it took nevertheless many days of hard study to supply the intermediate steps through which that transcendent mind had passed with one huge leap of inference. At some time in his youth no doubt Laplace had to think of these things, just as Rubinstein had once to think how his fingers should be placed on the keys of the piano; but what was once the object of conscious attention comes at last to be well-nigh automatic, while the night of the conscious mind goes on ever to higher and vaster themes.

Let us now take a long leap from the highest level of human intelligence to the mental life of a turtle or a codfish. In what does the mental life of such creatures consist? It consists of a few simple acts mostly concerned with the securing of food and the avoiding of danger, and these few simple acts are repeated with unvarying monotony during the whole lifetime of these creatures. Consequently these acts are performed with great ease and are attended with very little consciousness, and moreover the capacity to perform them is transmitted from parent to offspring as completely as the capacity of the stomach to digest food is transmitted. In all animals the new-born stomach needs but the contact with food in order to begin digesting, and the new-born lungs need but the contact with air in order to begin to breathe. The capacity for performing these perpetually repeated visceral actions is transmitted in perfection. All the requisite nervous connections are fully established during the brief embryonic existence of each creature. In the case of lower animals it is almost as much so with the few simple actions which make up the creature's mental life. The bird known as the fly-catcher no sooner breaks the egg than it will snap at and catch a fly. This action is not so very simple, but because it is something the bird is always doing, being indeed one out of the very few things that this bird ever does, the nervous connections needful for doing it are all established before birth, and nothing but the presence of the fly is required to set the operation going.

With such creatures as the codfish, the turtle, or the fly-catcher, there is accordingly nothing that can properly be called infancy. With them the sphere of education is extremely limited. They get their education before they are born. In other words, heredity does everything for them, education nothing. The career of the individual is predetermined by the careers of his ancestors, and he can do almost nothing to vary it. The life of such creatures is conservatism cut and dried, and there is nothing progressive about them.

In what I just said I left an "almost." There is a great deal of saving virtue in that little adverb. Doubtless even animals low in the scale possess some faint traces of educability; but they are so very slight that it takes geologic ages to produce an appreciable result. In all the innumerable wanderings, fights, upturnings and cataclysms of the earth's stupendous career, each creature has been summoned under penalty of death to use what little wit he may have had, and the slightest trace of mental flexibility is of such priceless value in the struggle for existence that natural selection must always have seized upon it, and sedulously hoarded and transmitted it for coming generations to strengthen and increase. With the lapse of geologic time the upper grades of animal intelligence have doubtless been raised higher and higher through natural selection. The warm-blooded mammals and birds of to-day no doubt surpass the cold-blooded dinosaurs of the Jurassic age in mental qualities as they surpass them in physical structure. From the codfish and turtle of ancient family to the modern lion, dog, and monkey, it is a very long step upward. The mental life of a warm-blooded animal is a very different affair from that of reptiles and fishes. A squirrel or a bear does a good many things in the course of his life. He meets various vicissitudes in various ways; he has adventures. The actions he performs are so complex and so numerous that they are severally performed with less frequency than the few actions performed by the codfish. The requisite nervous connections are accordingly not fully established before birth. There is not time enough. The nervous connections needed for the visceral movements and for the few simple instinctive actions get organized, and then the creature is born before he has learned how to do all the things his parents could do. A good many of his nervous connections are not yet formed, they are only formable. Accordingly he is not quite able to take care of himself; he must for a time be watched and nursed. All mammals and most birds have thus a period of babyhood that is not very long, but is on the whole longest with the most intelligent creatures. It is especially long with the higher monkeys, and among the man-like apes it becomes so long as to be strikingly suggestive. An infant orang-outang, captured by Mr. Wallace, was still a helpless baby at the age of three months, unable to feed itself, to walk without aid, or to grasp objects with precision.

But this period of helplessness has to be viewed under another aspect. It is a period of plasticity. The creature's career is no longer exclusively determined by heredity. There is a period after birth when its character can be slightly modified by what happens to it after birth, that is, by its experience as an individual. It becomes educable. It is no longer necessary for each generation to be exactly like that which has preceded. A door is opened through which the capacity for progress can enter. Horses and dogs, bears and elephants, parrots and monkeys, are all teachable to some extent, and we have even heard of a learned pig. Of learned asses there has been no lack in the world.

But this educability of the higher mammals and birds is after all quite limited. By the beginnings of infancy the door for progressiveness was set ajar, but it was not all at once thrown wide open. Conservatism stilt continued in fashion. One generation of cattle is much like another. It would be easy for foxes to learn to climb frees, and many a fox might have saved his life by doing so; yet quickwitted as he is, this obvious device never seems to have occurred to Reynard. Among slightly teachable mammals, however, there is one group more teachable than the rest. Monkeys, with their greater power of handling things, have also more inquisitiveness and more capacity for sustained attention than any other mammals; and the higher apes are fertile in varied resources. The orang-outang and gorilla are for this reason dreaded by other animals, and roam the undisputed lords of their native forests. They have probably approached the critical point where variations in intelligence, always important, have come to be supremely important, so as to be seized by natural selection in preference to variations in physical constitution. At some remote epoch of the past—we cannot say just when or how—our half-human forefathers reached and passed this critical point, and forthwith their varied struggles began age after age to result in the preservation of bigger and better brains, while the rest of their bodies changed but little. This particular work of natural selection must have gone on for an enormous length of time, and as its result we see that while man remains anatomically much like an ape, be has acquired a vastly greater brain with all that this implies. Zoologically the distance is small between man and the chimpanzee; psychologically it has become so great as to be immeasurable.

But this steady increase of intelligence, as our forefathers began to become human, carried with it a steady prolongation of infancy. As mental life became more complex and various, as the things to be learned kept ever multiplying, less and less could be done before birth, more and more must be left to be done in the earlier years of life. So instead of being born with a few simple capacities thoroughly organized, man came at last to be born with the germs of many complex capacities which were reserved to be unfolded and enhanced or checked and stifled by the incidents of personal experience in each individual. In this simple yet wonderful way there has been provided for man a long period during which his mind is plastic and malleable, and the length of this period has increased with civilization until it now covers nearly one third of our lives. It is not that our inherited tendencies and aptitudes are not still the main thing. It is only that we have at last acquired great power to modify them by training, so that progress may go on with ever-increasing sureness and rapidity.

In thus pointing out the causes of infancy, we have at the same time witnessed some of its effects. One effect, of stupendous importance, remains to be pointed out. As helpless babyhood came more and more to depend on parental care, the correlated feelings were developed on the part of parents, and the fleeting sexual relations established among mammals in general were gradually exchanged for permanent relations. A cow feels strong maternal affection for her nursing calf, but after the calf is fully grown, though doubtless she distinguishes it from other members of the herd, it is not clear that she entertains for it any parental feeling. But with our half-human forefathers it is not difficult to see how infancy extending over several years must have tended gradually to strengthen the relations of the children to the mother, and eventually to both parents, and thus give rise to the permanent organization of the family. When this step was accomplished we may say that the Creation of Man had been achieved. For through the organization of the family has arisen that of the clan or tribe, which has formed, as it were, the cellular tissue out of which the most complex human society has come to be constructed. And out of that subordination of individual desires to the common interest, which first received a definite direction when the family was formed, there grew the rude beginnings of human morality.

It was thus through the lengthening of his infancy that the highest of animals came to be Man,—a creature with definite social relationships and with an element of plasticity in his organization such as has come at last to make his difference from all other animals a difference in kind. Here at last there had come upon the scene a creature endowed with the capacity for progress, and a new chapter was thus opened in the history of creation. But it was not to be expected that man should all at once learn how to take advantage of this capacity. Nature, which is said to make no jumps, surely did not jump here. The whole history of civilization, indeed, is largely the history of man's awkward and stumbling efforts to avail himself of this flexibility of mental constitution with which God has endowed him. For many a weary age the progress men achieved was feeble and halting. Though it had ceased to be physically necessary for each generation to tread exactly in the steps of its predecessor, yet the circumstances of primitive society long made it very difficult for any deviation to be effected. For the tribes of primitive men were perpetually at war with each other, and their methods of tribal discipline were military methods. To allow much freedom of thought would be perilous, and the whole tribe was supposed to be responsible for the words and deeds of each of its members. The tribes most rigorous in this stern discipline were those which killed out tribes more loosely organized, and thus survived to hand down to coming generations their ideas and their methods. From this state of things an intense social conservatism was begotten,—a strong disposition on the part of society to destroy the flexible-minded individual who dares to think and behave differently from his fellows. During the past three thousand years much has been done to weaken this conservatism by putting an end to the state of things which produced it. As great and strong societies have arisen, as the sphere of warfare has diminished while the sphere of industry has enlarged, the need for absolute conformity has ceased to be felt, while the advantages of freedom and variety come to be ever more clearly apparent. At a late stage of civilization, the flexible or plastic society acquires even a military advantage over the society that is more rigid, as in the struggle between French and English civilization for primacy in the world. In our own country, the political birth of which dates from the triumph of England in that mighty struggle, the element of plasticity in man's nature is more thoroughly heeded, more fully taken account of, than in any other community known to history; and herein lies the chief potency of our promise for the future. We have come to the point where we are beginning to see that we may safely depart from unreasoning routine, and, with perfect freedom of thinking in science and in religion, with new methods of education that shall train our children to think for themselves while they interrogate Nature with a courage and an insight that shall grow ever bolder and keener, we may ere long be able fully to avail ourselves of the fact that we come into the world as little children with undeveloped powers wherein lie latent all the boundless possibilities of a higher and grander Humanity than has yet been seen upon the earth.

II

THE PART PLAYED BY INFANCY IN THE EVOLUTION OF MAN

The remarks which my friend Mr. Clark has made with reference to the reconciling of science and religion seem to carry me back to the days when I first became acquainted with the fact that there were such things afloat in the world as speculations about the origin of man from lower forms of life; and I can recall step by step various stages in which that old question has come to have a different look from what it had thirty years ago. One of the commonest objections we used to hear, from the mouths of persons who could not very well give voice to any other objection, was that anybody, whether he knows much or little about evolution, must have the feeling that there is something degrading about being allied with lower forms of life. That was, I suppose, owing to the survival of the old feeling that a dignified product of creation ought to have been produced in some exceptional way. That which was done in the ordinary way, that which was done through ordinary processes of causation, seemed to be cheapened and to lose its value. It was a remnant of the old state of feeling which took pleasure in miracles, which seemed to think that the object of thought was more dignified if you could connect it with something supernatural; that state of culture in which there was an altogether inadequate appreciation of the amount of grandeur that there might be in the slow creative work that goes on noiselessly by little minute increments, even as the dropping of the water that wears away the stone. The general progress of familiarity with the conception of evolution has done a great deal to change that state of mind. Even persons who have not much acquaintance with science have at length caught something of its lesson,—that the infinitely cumulative action of small causes like those which we know is capable of producing results of the grandest and most thrilling importance, and that the disposition to recur to the cataclysmic and miraculous is only a tendency of the childish mind which we are outgrowing with wider experience.

The whole doctrine of evolution, and in fact the whole advance of modern science from the days of Copernicus down to the present day, have consisted in the substitution of processes which are familiar and the application of those processes, showing how they produce great results.

When Darwin's "Origin of Species" was first published, when it gave us that wonderful explanation of the origin of forms of life from allied forms through the operation of natural selection, it must have been like a mental illumination to every person who comprehended it. But after all it left a great many questions unexplained, as was natural. It accounted for the phenomena of organic development in general with wonderful success, but it must have left a great many minds with the feeling: If man has been produced in this way, if the mere operation of natural selection has produced the human race, wherein is the human race anyway essentially different from lower races? Is not man really dethroned, taken down from that exceptional position in which we have been accustomed to place him, and might it not be possible, in the course of the future, for other beings to come upon the earth as far superior to man as man is superior to the fossilized dragons of Jurassic antiquity?

Such questions used to be asked, and when they were asked, although one might have a very strong feeling that it was not so, at the same time one could not exactly say why. One could not then find any scientific argument for objections to that point of view. But with the further development of the question the whole subject began gradually to wear a different appearance; and I am going to give you a little bit of autobiography, because I think it may be of some interest in this connection. I am going to mention two or three of the successive stages which the whole question took in my own mind as one thing came up after another, and how from time to time it began to dawn upon me that I had up to that point been looking at the problem from not exactly the right point of view.

When Darwin's "Descent of Man" was published in 1871, it was of course a book characterized by all his immense learning, his wonderful fairness of spirit and fertility of suggestion. Still, one could not but feel that it did not solve the question of the origin of man. There was one great contrast between that book and his "Origin of Species." In the earlier treatise he undertook to point out a *vera causa* of the origin of species, and he did it. In his "Descent of Man" he brought together a great many minor generalizations which facilitated the understanding of man's origin. But he did not come at all near to solving the central problem, nor did he anywhere show clearly why natural selection might not have gone on forever producing one set of beings after another distinguishable chiefly by physical differences. But Darwin's co-discoverer, Alfred Russel Wallace, at an early stage in his researches, struck out a most brilliant and pregnant suggestion. In that one respect Wallace went further than ever Darwin did. It was a point of which, indeed, Darwin admitted the importance. It was a point of which nobody could fail to understand the importance, that in the course of the evolution of a very highly organized animal, if there came a point at which it was of more advantage to that animal to have variations in his intelligence seized upon and improved by natural selection than to have physical changes seized upon, then natural selection would begin working almost exclusively upon that creature's intelligence, and he would develop in intelligence to a great extent, while his physical organism would change but slightly. Now, that of course applied to the case of man, who is changed physically but very slightly from the apes, while he has traversed intellectually such a stupendous chasm.

As soon as this statement was made by Wallace, it seemed to me to open up an entirely new world of speculation. There was this enormous antiquity of man, during the greater part of which he did not know enough to make history. We see man existing here on the earth, no one can say how long, but surely many hundreds of thousands of years, yet only during just the last little fringe of four or five thousand years has he arrived at the point where he makes history. Before that, something was going on, a great many things were going on, while his ancestors were slowly growing up to that point of intelligence where it began to make itself felt in the recording of events. This agrees with Wallace's suggestion of a long period of psychical change, accompanied by slight physical change.

Well, in the spring of 1871, when Darwin's "Descent of Man" came out, just about the same time I happened to be reading Wallace's account of his experiences in the Malay Archipelago, and how at one time he caught a female orang-outang with a new-born baby, and the mother died, and Wallace brought up the baby orang-outang by hand; and this baby orang-outang had a kind of infancy which was a great deal longer than that of a cow or a sheep, but it was nothing compared to human infancy in length. This little orang-outang could not get up and march around, as mammals of less intelligence do, when he was first born, or within three or four days; but after three or four weeks or so he would get up, and begin taking hold of something and pushing it around, just as children push a chair; and he went through a period of staring at his hands, as human babies do, and altogether was a good deal slower in getting to the point where he could take care of himself. And while I was reading of that I thought, Dear me! if there is any one thing in which the human race is signally distinguished from other mammals, it is in the enormous duration of their infancy; but it is a point that I do not recollect ever seeing any naturalist so much as allude to.

It happened at just that time that I was making researches in psychology about the organization of experiences, the way in which conscious intelligent action can pass down into quasi-automatic action, the generation of instincts, and various allied questions; and I thought, Can it be that the increase of intelligence in an animal, if carried beyond a certain point, must necessarily result in prolongation of the period of infancy,—must necessarily result in the birth of the mammal at a less developed stage, leaving something to be done, leaving a good deal to be done, after birth? And then the argument seemed to come along very naturally, that for every action of life, every adjustment which a creature makes in life, whether a muscular adjustment or an intelligent adjustment, there has got to be some registration effected in the nervous system, some line of transit worn for nervous force to follow; there has got to be a connection between certain nerve-centres before the thing can be done, whether it is the acts of the viscera or the acts of the limbs, or anything of that sort; and of course it is obvious that if the creature has not many things to register in his nervous system, if he has a life which is very simple, consisting of few actions that are performed with great frequency, that animal becomes almost automatic in his whole life; and all the nervous connections that need to be made to enable him to carry on life get made during the foetal period or during the egg period, and when he comes to be born, he comes all ready to go to work. As one result of this, he does not learn from individual experience, but one generation is like the preceding generations, with here and there some slight modifications. But when you get the creature that has arrived at the point where his experience has become varied, he has got to do a good many things, and there is more or less individuality about them; and many of them are not performed with the same minuteness and regularity, so that there does not begin to be that automatism within the period during which he is being developed and his form is taking on its outlines. During prenatal life there is not time enough for all these nervous registrations, and so by degrees it comes about that he is born with his nervous system perfectly capable only of making him breathe and digest food,—of making him do the things absolutely requisite for supporting life; instead of being born with a certain number of definite developed capacities, he has a number of potentialities which have got to be roused according to his own individual experience. Pursuing that line of thought, it began after a while to seem clear to me that the infancy of the animal in a very undeveloped condition, with the larger part of his faculties in potentiality rather than in actuality, was a direct result of the increase of intelligence, and I began to see that now we have two steps: first, natural selection goes on increasing the intelligence; and secondly, when the intelligence goes far enough, it makes a longer infancy, a creature is born less developed, and therefore there comes this plastic period during which he is more teachable. The capacity for progress begins to come in, and you begin to get at one of the great points in which man is distinguished from the lower animals, for one of those points is undoubtedly his progressiveness; and I think that any one will say, with very little hesitation, that if it were not for our period of infancy we should not be progressive. If we came into the world with our capacities all cut and dried, one generation would be very much like another.

Then, looking round to see what are the other points which are most important in which man differs from the lower animals, there comes that matter of the family. The family has adumbrations and foreshadowings among the lower animals, but in general it may be said that while mammals lower than man are gregarious, in man have become established those peculiar relationships which constitute what we know as the family; and it is easy to see how the existence of helpless infants would bring about just that state of things. The necessity of caring for the infants would prolong the period of maternal affection, and would tend to keep the father and mother and children together, but it would tend especially to keep the mother and children together. This business of the marital relations was not really a thing that became adjusted in the primitive ages of man, but it has become adjusted in the course of civilization. Real monogamy, real faithfulness of the male parent, belongs to a comparatively advanced stage; but in the early stages the knitting together of permanent relations between mother and infant, and the approximation toward steady relations on the part of the male parent, came to bring about the family, and gradually to knit those organizations which we know as clans.

Here we come to another stage, another step forward. The instant society becomes organized in clans, natural selection cannot let these clans be broken up and die out,—the clan becomes the chief object or care of natural selection, because if you destroy it you retrograde again, you lose all you have gained; consequently, those clans in which the primeval selfish instincts were so modified that the individual conduct would be subordinated to some extent to the needs of the clan,—those are the ones which would prevail in the struggle for life. In this way you gradually get an external standard to which man has to conform his conduct, and you get the germs of altruism and morality; and in the prolonged affectionate relation between the mother and the infant you get the opportunity for that development of altruistic feeling which, once started in those relations, comes into play in the more general relations, and makes more feasible and more workable the bonds which keep society together, and enable it to unite on wider and wider terms.

So it seems that from a very small beginning we are reaching a very considerable result. I had got these facts pretty clearly worked out, and carried them around with me some years, before a, fresh conclusion came over me one day with a feeling of surprise. In the old days before the Copernican astronomy was promulgated, man regarded himself as the centre of the universe. He used to entertain theological systems which conformed to his limited knowledge of nature. The universe seemed to be made for his uses, the earth seemed to have been fitted up for his dwelling place, he occupied the centre of creation, the sun was made to give him light, etc. When Copernicus overthrew that view, the effect upon theology was certainly tremendous. I do not believe that justice has ever been done to the shock that it gave to man when he was made to realize that he occupied a kind of miserable little clod of dirt in the universe, and that there were so many other worlds greater than this. It was one of the first great shocks involved in the change from ancient to modern scientific views, and I do not doubt it was responsible for a great deal of the pessimistic philosophizing that came in the seventeenth and eighteenth centuries.

Now, it flashed upon me a dozen years or so ago—after thinking about this manner in which man originated—that man occupies certainly just as exceptional a position as before, if he is the terminal in a long series of evolutionary events. If at the end of the long history of evolution comes man, if this whole secular process has been going on to produce this supreme object, it does not much matter what kind of a cosmical body he lives on. He is put back into the old position of theological importance, and in a much more intelligent way than in the old days when he was supposed to occupy the centre of the universe. We are enabled to say that while there is no doubt of the evolutionary process going on throughout countless ages which we know nothing about, yet in the one case where it is brought home to us we spell out an intelligible story, and we do find things working along up to man as a terminal fact in the whole process. This is indeed a consistent conclusion from Wallace's suggestion that natural selection, in working toward the genesis of man, began to follow a new path and make psychical changes instead of physical changes. Obviously, here you are started upon a new chapter in the history of the universe. It is no longer going to be necessary to shape new limbs, and to thicken the skin and make new growths of hair, when man has learned how to build a fire, when he can take some other animal's hide and make it into clothes. You have got to a new state of things.

After I had put together all these additional circumstances with regard to the origination of human society and the development of altruism, I began to see a little further into the matter. It then began to appear that not only is man the terminal factor in a long process of evolution, but in the origination of man there began the development of the higher psychical attributes, and those attributes are coming to play a greater and greater part in the development of the human race. Just take this mere matter of "altruism," as we call it. It is not a pretty word, but must serve for want of a better. In the development of altruism from the low point, where there was scarcely enough to hold the clan together, up to the point reached at the present day, there has been a notable progress, but there is still room for an enormous amount of improvement. The progress has been all in the direction of bringing out what we call the higher spiritual attributes. The feeling was now more strongly impressed upon me than ever, that all these things tended to set the whole doctrine of evolution into harmony with religion; that if the past through which man had originated was such as has been described, then religion was a fit and worthy occupation for man, and some of the assumptions which underlie every system of religion must be true. For example, with regard to the assumption that what we see of the present life is not the whole thing; that there is a spiritual side of the question beside the material side; that, in short, there is for man a life eternal. When I wrote the "Destiny of Man," all that I ventured to say was, that it did not seem quite compatible with ordinary common sense to suppose that so much pains would have been taken to produce a merely ephemeral result. But since then another argument has occurred to me: that just at the time when the human race was beginning to come upon the scene, when the germs of morality were coming in with the family, when society was taking its first start, there came into the human mind—how one can hardly say, but there did come—the beginnings of a groping after something that lies outside and beyond the world of sense. That groping after a spiritual world has been going on here for much more than a hundred thousand years, and it has played an enormous part in the history of mankind, in the whole development of human society. Nobody can imagine what mankind would have been without it up to the present time. Either all religion has been a reaching out for a phantom that does not exist, or a reaching out after something that does exist, but of which man, with his limited intelligence, has only been able to gain a crude idea. And the latter seems a far more probable conclusion, because, if it is not so, it constitutes a unique exception to all the operations of evolution we know about. As a general thing in the whole history of evolution, when you see any internal adjustment reaching out toward something, it is in order to adapt itself to something that really exists; and if the religious cravings of man constitute an exception, they are the one thing in the whole process of evolution that is exceptional and different from all the rest. And this is surely an argument of stupendous and resistless weight.

I take this autobiographical way of referring to these things, in the order in which they came before my mind, for the sake of illustration. The net result of the whole is to put evolution in harmony with religious thought,—not necessarily in harmony with particular religious dogmas or theories, but in harmony with the great religious drift, so that the antagonism which used to appear to exist between religion and science is likely to disappear. So I think it will before a great while. If you take the case of some evolutionist like Professor Haeckel, who is perfectly sure that materialism accounts for everything (he has got it all cut and dried and settled; he knows all about it, so that there is really no need of discussing the subject!); if you ask the question whether it was his scientific study of evolution that really led him to such a dogmatic conclusion, or whether it was that he started from some purely arbitrary assumption, like the French materialists of the eighteenth century, I have no doubt the latter would be the true explanation. There are a good many people who start on their theories of evolution with these ultimate questions all settled to begin with. It was the most natural thing in the world that after the first assaults of science upon old beliefs, after a certain number of Bible stories and a certain number of church doctrines had been discredited, there should be a school of men who in sheer weariness should settle down to scientific researches, and say, "We content ourselves with what we can prove by the methods of physical science, and we will throw everything else overboard." That was very much the state of mind of the famous French atheists of the last century. But only think how chaotic nature was to their minds compared to what she is to our minds to-day. Just think how we have in the present century arrived where we can see the bearings of one set of facts in nature as collated with another set of facts, and contrast it with the view which even the greatest of those scientific French materialists could take. Consider how fragmentary and how lacking in arrangement was the universe they saw compared with the universe we can see to-day, and it is not strange that to them it could be an atheistic world. That hostility between science and religion continued as long as religion was linked hand in hand with the ancient doctrine of special creation. But now that the religious world has unmoored itself, now that it is beginning to see the truth and beauty of natural science and to look with friendship upon conceptions of evolution, I suspect that this temporary antagonism, which we have fallen into a careless way of regarding as an everlasting antagonism, will come to an end perhaps quicker than we realize.

There is one point that is of great interest in this connection, although I can only hint at it. Among the things that happened in that dim past when man was coming into existence was the increase of his powers of manipulation; and that was a factor of immense importance. Anaxagoras, it is said, wrote a treatise in which he maintained that the human race would never have become human if it had not been for the hand. I do not know that there was so very much exaggeration about that. It was certainly of great significance that the particular race of mammals whose intelligence increased far enough to make it worth while for natural selection to work upon intelligence alone was the race which had developed hands and could manipulate things. It was a wonderful era in the history of creation when that creature could take a club and use it for a hammer, or could pry up a stone with a stake, thus adding one more lever to the levers that made up his arm. From that day to this, the career of man has been that of a person who has operated upon his environment in a different way from any animal before him. An era of similar importance came probably somewhat later, when man learned how to build a fire and cook his food; thus initiating that course of culinary development of which we have seen the climax in our dainty dinner this evening. Here was another means of acting upon the environment. Here was the beginning of the working of endless physical and chemical changes through the application of heat, just as the first use of the club or the crowbar was the beginning of an enormous development in the mechanical arts.

Now, at the same time, to go back once more into that dim past, when ethics and religion, manual art and scientific thought, found expression in the crudest form of myths, the aesthetic sense was germinating likewise. Away back in the glacial period you find pictures drawn and scratched upon the reindeer's antler, portraitures of mammoths and primitive pictures of the chase; you see the trinkets, the personal decorations, proving beyond question that the aesthetic sense was there. There has been an immense aesthetic development since then. And I believe that in the future it is going to mean far more to us than we have yet begun to realize. I refer to the kind of training that comes to mankind through direct operation upon his environment, the incarnation of his thought, the putting of his ideas into new material relations. This is going to exert powerful effects of a civilizing kind. There is something strongly educational and disciplinary in the mere dealing with matter, whether it be in the manual training school, whether it be in carpentry, in overcoming the inherent and total depravity of inanimate things, shaping them to your will, and also in learning to subject yourself to their will (for sometimes you must do that in order to achieve your conquests; in other words, you must humour their habits and proclivities). In all this there is a priceless discipline, moral as well as mental, let alone the fact that, in whatever kind of artistic work a man does, he is doing that which in the very working has in it an element of something outside of egoism; even if he is doing it for motives not very altruistic, he is working toward a result the end of which is the gratification or the benefit of other persons than himself; he is working toward some result which in a measure depends upon their approval, and to that extent tends to bring him into closer relations to his fellow man.

In the future, to an even greater extent than in the recent past, crude labour will be replaced by mechanical contrivances. The kind of labour which can command its price is the kind which has trained intelligence behind it. One of the great needs of our time is the multiplication of skilled and special labour. The demand for the products of intelligence is far greater than that for mere crude products of labour, and it will be more and more so. For there comes a time when the latter products have satisfied the limit to which a man can consume food and drink and shelter,—those things which merely keep the animal alive. But to those things which minister to the requirements of the spiritual side of a man, there is almost no limit. The demand one can conceive is well-nigh infinite. One of the philosophical things that have been said, in discriminating man from the lower animals, is that he is the one creature who is never satisfied. It is well for him that he is so, that there is always something more for which he craves. To my mind, this fact most strongly hints that man is infinitely more than a mere animate machine.

OUTLINE

I. THE MEANING OF INFANCY

1. The relation between progress and infancy 2. Man's method of learning 3. The mental inheritance of animals 4. Infancy and educability of animals 5. Infancy is a period of plasticity 6. Educability varies widely in different creatures 7. Increased intelligence means prolonged infancy 8. The socializing effects of infancy 9. The use of this capacity for progress in the past

II. THE PART PLAYED BY INFANCY IN THE EVOLUTION OF MAN

1. The grandeur of natural causation
2. The problem of man's ascendancy
3. Natural selection seizes on intelligence
4. A long infancy characteristic of man
5. A complex life requires a longer infancy
6. Infancy fosters sociability and the family
7. Group life increases the social and moral bonds
8. Spiritual man is evolution's terminal factor
9. Man marks a development along new lines
10. Hand-work in the evolution of intelligence
11. The educational value of aesthetic effort
12. Man's spirituality is prophetic of his destiny

CPSIA information can be obtained
at www.ICGtesting.com
Printed in the USA
LVHW041023020323
740706LV00008B/1144

9 789356 895850